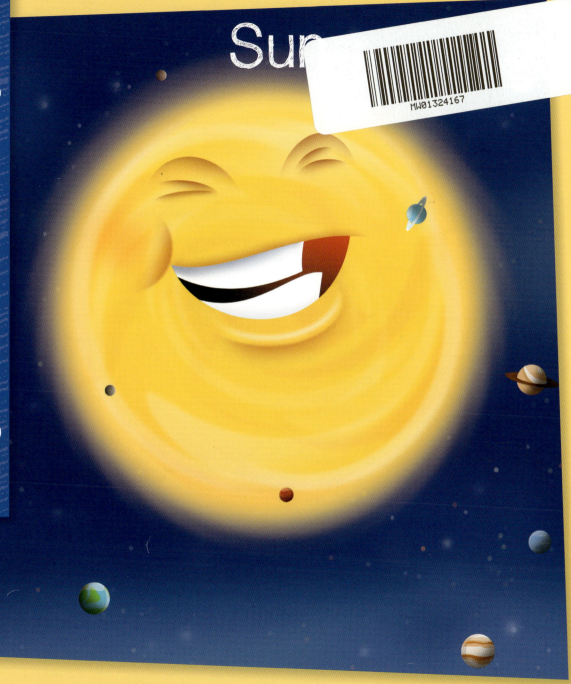

Sun

my guide to the solar system

CHERRY LAKE PRESS

Published in the United States of America by Cherry Lake Publishing
Ann Arbor, Michigan
www.cherrylakepublishing.com

Reading Adviser: Beth Walker Gambro, MS, Ed., Reading Consultant, Yorkville, IL
Book Design: Jennifer Wahi
Illustrator: Jeff Bane

Photo Credits: © Beautiful landscape/Shutterstock.com, 5; © ClaudioVentrella/iStock.com, 7; © DKosig/iStock.com, 9; © libre de droit/iStock.com, 11; © GSFC_20171208/NASA, 13; © Skylines/Shutterstock.com, 15; © Outer Space/Shutterstock.com, 17; © dzika_mrowka/iStock.com, 19; © Vadim Sadovski/Shutterstock.com, 21; © aryos/iStock.com, 23; Cover, 2-3, 6, 18, 22, 24, Jeff Bane

Copyright © 2022 by Cherry Lake Publishing Group
All rights reserved. No part of this book may be reproduced or utilized in any form or by any means without written permission from the publisher.

Cherry Lake Press is an imprint of Cherry Lake Publishing Group.

Library of Congress Cataloging-in-Publication Data

Names: Devera, Czeena, author. | Bane, Jeff, 1957- illustrator.
Title: Sun / by Czeena Devera ; illustrated by Jeff Bane.
Description: Ann Arbor, Michigan : Cherry Lake Publishing, [2022] | Series: My guide to the solar system | Audience: Grades K-1
Identifiers: LCCN 2021036762 (print) | LCCN 2021036763 (ebook) | ISBN 9781534199033 (hardcover) | ISBN 9781668900178 (paperback) | ISBN 9781668905937 (ebook) | ISBN 9781668901618 (pdf)
Subjects: LCSH: Sun--Juvenile literature.
Classification: LCC QB521.5 .D48 2023 (print) | LCC QB521.5 (ebook) | DDC 523.7--dc23
LC record available at https://lccn.loc.gov/2021036762
LC ebook record available at https://lccn.loc.gov/2021036763

Printed in the United States of America
Corporate Graphics

table of contents

Sun 4

Glossary 24

Index 24

About the author: Czeena Devera grew up in the red-hot heat of Arizona surrounded by books. Her childhood bedroom had built-in bookshelves that were always full. She now lives in Michigan with an even bigger library of books.

About the illustrator: Jeff Bane and his two business partners own a studio along the American River in Folsom, California, home of the 1849 Gold Rush. When Jeff's not sketching or illustrating for clients, he's either swimming or kayaking in the river to relax.

Sun

I'm the Sun. You see me every day! Unless the sky is cloudy.

I'm about 109 times bigger than Earth.

My outer layer is called the corona. You can see it during a **solar eclipse**.

My inner layer is called the core. This is where I get my heat and light.

I'm made up of very hot gases. They're so hot that they're **plasma**.

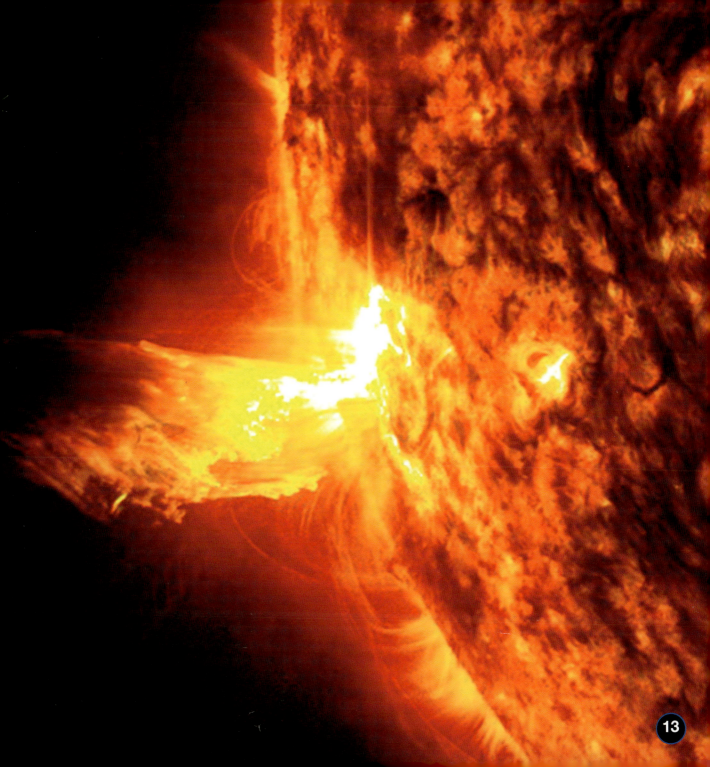

I'm at the center of our **solar system**. The planets **orbit** around me.

The solar system and I orbit the Milky Way **galaxy**. It takes me about 250 million years to complete my orbit!

Stars come in different colors and sizes. I'm part of a group called yellow **dwarf** stars.

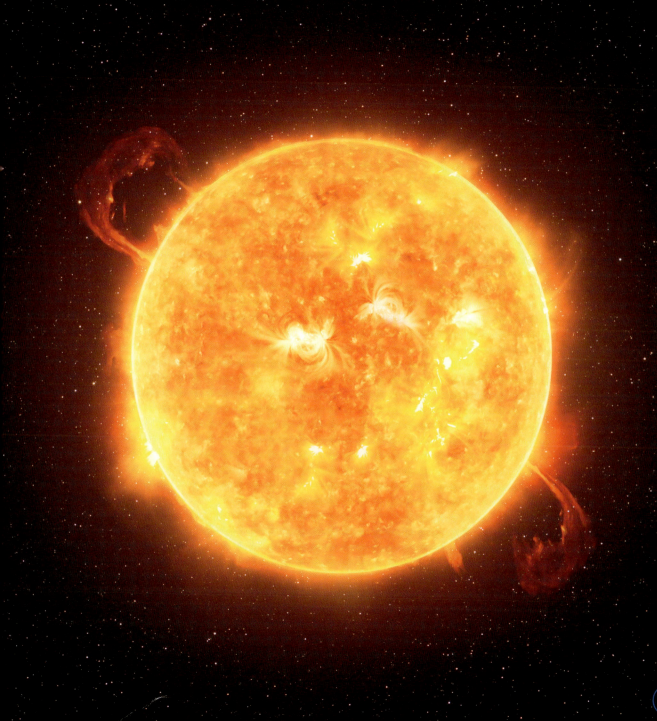

I'm young for a star. I'm only 4.5 billion years old.

Scientists are still studying me. There's so much more to learn!

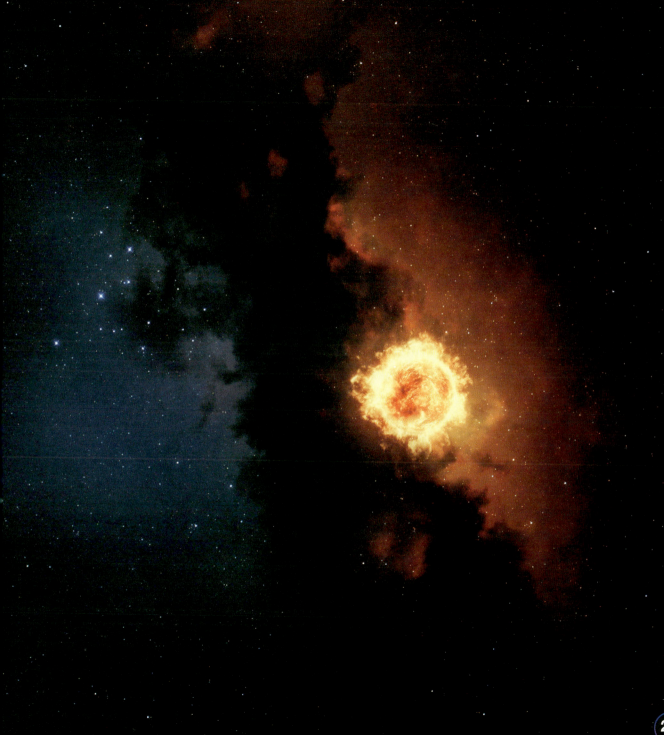

glossary & index

glossary

dwarf (DWORF) a smaller star, like Earth's Sun

galaxy (GAH-luhk-see) one of the large groups of stars that make up the universe

orbit (OR-buht) to travel in a curved path around something, such as a planet or star

plasma (PLAZ-muh) a state of matter similar in some ways to a gas

scientists (SYE-uhn-tists) people who study nature and the world we live in

solar eclipse (SOH-luhr i-KLIPS) when the Sun is hidden by the Moon

solar system (SOH-luhr SIH-stuhm) a star and the planets that move around it

index

core, 10
corona, 8

Earth, 6

Milky Way, 16

orbit, 14, 16

solar system, 14, 16
stars, 18, 20